Living In Clown World – The Trickster Zeitgeist...

Copyright © 2023 Wayne McRoy

All Rights Reserved

ISBN: 9798866536214

You're listening to the Alchemical Tech Revolution and I am your host.

Wayne McRoy.

Good evening, everyone.

Tonight, we're going to take a look at what it's like living in clown world, the trickster Zeitgeist.

Have you had a feeling as of late that we live in clown world?

5:40

Have you seen the memes?

Relating to that idea.

The fact that we live in clown world.

Have you heard people refer to it as clown World?

Well, I assure you folks.

They're not far from the mark because we have been living in clown world and I could tell you precisely when clown World actually launched as a very real thing.

6:02

Here, it's an archetype, it's being used and employed by the social controllers of This World.

By the ones, pulling the strings at the top of the power structure.

These dark occultists who run things, and they are laughing at you and mocking you with this trickster Zeitgeist that they have effectuated in the Modern Age now.

6:25

So tonight we're going to explore that topic and we're going to put make try to make a little sense out of what's happened in the world over the course of the past several years.

And we'll see as we explore this.

That this goes much deeper than just the idea of say clowns in a circus and all the very silly things that are associated with the Clown World meme.

6:53

Because this is a very real meme, okay, this is the alchemical meme put into full effect here in the times.

We are living in and this has been done.

On a very Grand scale, and we are living in an era where the people in charge are fools.

7:17

Not only fools, but they're incompetent, their Jesters, they're tricksters, the trickster archetype as it were the Zeitgeist of the trickster, the Zeitgeist of the clown.

And we'll explore this Avenue of thought.

Tonight, as we read into a book here that actually came from Yale, at University Library in 1956 book written by Mr.

7:42

Paul Raiden called "the trickster a study in American Indian mythology", and we're going to read a portion of this tonight because this has commentaries and contributions from Karl Kerenyi and Carl, G Jung.

And we're going to read the portion of the book here by Carl Jung and this Part is called "on the psychology of the trickster figure".

8:07

And we'll see as we explore into this that we've talked about some of these things in past episodes here.

We've taken notice to some of these things.

We've pointed them out and nauseam.

And when I tell you, we're living in clown, world.

8:23

I'm not kidding.

We truly are.

This has been inaugurated in our world in our Western culture here and it's something that's being held up.

And there's a very real spiritual connotation behind all of it.

8:39

There's a very real energy behind this in our You type, if you will, that affects the human mind on many levels.

And as we delve into the works, here of Carl Jung we will see exactly what we're talking about because you might be surprised to find out that Carl Jung is talking about.

9:00

Well, some very strange occultic, things that tie back to older times.

So that being the case, we could rest assured that these darker cultists who run things in this world very much, have a firm grasp of these ideas as well.

9:18

And have most likely purposely, put this archetype into play in this age, created this modern clown World.

Zeitgeist that we currently have and we'll see as we get through the reading here, just the corresponding evidence.

9:37

That lays this all down for your modern mind here.

So let's get right into it.

Without further Ado, on the psychology of the trickster figure, it is no light task for me to write about the figure of the trickster in American Indian mythology within the confined space of a commentary.

9:57

When I first came across Adolf Bandoliers classic on this subject,

"The Delight Makers", many years ago, I was struck by the European analogy of the tradition in the medieval church with its reversal of the hierarchical order, which is still continued in the Carnival's held by student societies today, something of this contradictory ethos also in here is in the medieval description of the devil as simia dei, the ape of God, and in his characterization, in folklore, as the simpleton who is fooled or cheated, a curious combination of typical trickster motifs, can be found in the alchemical Figure of mere curious.

10:39

For instance, his fondness for slide jokes and malicious pranks his powers as a shapeshifter, his dual nature half animal, half Divine his exposure, to all kinds of tortures.

And last, but not least his approximation to the figure of a savior.

10:57

These qualities make mercurius seem like a demonic being resurrected from primitive times older even than the Greek Hermes his roguery. he's relate him in some measure, to various big ears met with in folklore and universally known in fairy tales Tom Thumb stupid Hans?

11:17

Or the buffoon like Hans worst who isn't altogether - hero.

And yet manages to achieve though, or through his stupidity, what others fail to accomplish with their best efforts.

In Grimm's, fairytale the spirit mercurius, lets himself be outwitted by a peasant lab and then has to buy his freedom with the precious gift of healing.

11:41

And I'm going to pause for a second right there.

Folks.

So pointing out earlier, Leon here.

Carl Jung noticed that the European analogy of the carnival in the medieval church with its reversal of the hierarchical order.

11:58

Was a perfect.

Representation of this trickster figure or clown Motif.

He's referring, of course, to a subject.

We've covered here on past broadcast the Feast of fools.

12:16

Also, the Feast of epiphany, which the Feast of fools was celebrated on the date of epiphany every year. so, he noticed something, That we've seen here and we've discussed already on several occasions.

12:37

Now, when I can't stress the importance of this enough because this is something that is a hugely important event.

That will shape our future here and as actually inaugurated us into the clown World Motif, so to say here, but let's continue reading what Young has to say.

13:04

And we'll pontificate more on that in a little while here since all mythical figures correspond to Inner psychic experiences and originally sprang from them, it is not surprising to find certain phenomena in the field of parapsychology.

13:20

Which reminds us of the trickster.

These are the phenomena connected with poltergeists and they occur.

All at all times and places in the ambience of pre-adolescent children.

The malicious tricks played by the Poltergeist are as well known as the low level of his intelligence and the fatuity of his Communications ability to change.

13:42

His shape, seems almost to be one of his characteristics.

As there are not a few reports of his appearance in animal form since Has on occasion described himself as a soul in Hell.

The motif of subjective suffering.

Would seem not to be lacking either, his universality is coextensive.

14:01

So, to speak with that of shamanism to which, as we know the whole phenomenology of spiritualism belongs, there is something of the trickster in the character of the shaman, in Medicine, Man.

For he too often plays malicious jokes on people only to fall victim in his turn.

14:20

Earn to the Vengeance of Those whom he has injured for this reason his profession sometimes puts him in Peril of his life.

Besides that, the shamanistic techniques in themselves often, cause the medicine, man, a good deal of discomfort, if not actual pain at all events, the making of a medicine man involves in many parts of the world so much Agony of body and soul that permanent psychic injuries.

14:45

May result his approximation to the savior is a no Obvious consequence of this in confirmation of the mythological truth that the wounded wounder is the agent of healing and that the sufferer takes away suffering in a pause.

15:03

There for a moment, folks.

So think about what Young just said there about this archetype, you see and relate it to some things that we know.

And we've seen this whole idea of mercury or Hermes being the trickster also known by other, Appalachians to Prometheus, same figure described in different ways.

15:32

Loki in the Norse mythology, the trickster.

Let's continue reading these mythological features, extend even to the highest regions of man's Spiritual Development.

If we consider for example, the Demonic features exhibited by Yahweh in the Old Testament, we shall find in them, not a few reminders of the unpredictable behavior of the trickster of his pointless.

15:56

Orgies of Destruction in his self-appointed, sufferings together with the same gradual development into a savior and his simultaneous humanization It is just this transformation of the meaningless into the meaningful that reveals the tricksters compensatory relation to the Saint which in the Early Middle Ages.

16:16

Led to some strange ecclesiastical Customs based on memories of the ancient saturnalia mostly they were celebrated on the days.

Immediately following the birth of Christ that is, in the new year, with singing and dancing.

16:33

And I'm going to pause for a moment here, folks, in the New Year, he's most definitely referring to the Feast of fools as it were.

He's referring to the date that we now know in infamy here, January 6th 2021 is when clown world was thrust Upon Us in full here.

16:55

This archetype leveraged against us, invoking, the Feast of fools this old call back to saturnalia, saturnalia celebrations.

And he goes into more detail as we go a little further here, but we could also see he's talking about the idea of the days immediately following the birth of Christ.

17:16

And this was the New Year back in the medieval times.

And what is also heavily being portrayed in the news over the course of the past several days.

Now about the past week, week and a half, Well they're talking about the Lunar New Year, the Chinese New Year, the year of the rabbit, right?

17:37

So we see all of these things being invoked here and it all has to do with this trickster archetype because the rabbit is also seen as a trickster, isn't he?

He's clever and he likes to play tricks upon his enemies or even his friends.

17:58

Look at for example, the Tune character Bugs, Bunny, what is he known for?

Well, he's got a quick wit and he does play tricks.

He's got this trickster archetype to him as well.

So, there's an invocation going on of this trickster archetype and mass here relating to the new year and the celebration of the Feast of epiphany back on.

18:22

January 6th, 2021 with the call, back to this Feast of fools that we've discussed on past programs.

A, if you haven't listened to those, please go back and take a listen, because this is important.

And the more that I have been exploring this topic, the more of this kind of stuff, I'm finding.

18:41

And apparently, it's not just me, that's noticing this archetype here.

Because Carl Jung pointed out this archetype is being hugely important here.

Somebody's leveraging, this archetype folks.

Somebody at the top of the power structure that's looking to achieve certain goals.

19:00

They've set this thing in motion.

They've invoked this new Zeitgeist into the modern day.

Here, clown world is in full swing.

It started January 6, 20 21 C.

19:16

So this was something hugely important, but not for the reasons that are presented to you in media.

This was obviously a mockery, this whole situation.

A mockery as we'll see here, but let's continue on with the reading here and see what else Carl Jung has to say about this, the dances were originally harmless, triperdia of the priests, the lower clergy children and sub deacons and they took place in the church an episcopis purirum.

19:48

The Children's Bishop was elected and dressed in pontifical robes, amid uproarious rejoicing.

He paid An official visit to the Palace of the Archbishop and distributed the Episcopal blessing from one of the windows.

The same thing happened at the tripudeium hypo dianorum and at the dances, for other Priestly grades, by the end of the 12th, century of the sub-deacon's

20:14

dance had already.

Degenerated into a Festum stulurum, a Fool's Feast . A report from the year 1198 says that the Feast of circumcision in Notre Dame

Paris so many Abominations and shameful Deeds were committed that the holy place was desecrated, not only by smutty jokes.

20:34

But even by the shedding of blood in vain did Pope Innocent the third inveigh against the jests and Madness that make the clergy a mockery and the Shameless frenzy of their

Play-acting nearly 300 years later, the 12th of March 14, 4400 letter from the theological faculty of Paris, to all the French Bishops, Was still fulminating against these festivals at which even the priests and clerics elected in Archbishop or a bishop or a pope, and named him.

21:06

The fools Pope Futurum Popham, and I'm going to pause for a moment here.

Folks, they elected a Fool's Pope who was elected, In 2020.

21:26

And was inaugurated in January of twenty Twenty-One just weeks or days even after this January, 6th 2021 Feast of fools.

Well, the fool himself, they put in a Fool's Pope In the position of leadership.

21:51

Do you think Joe Biden's up there by accident?

Do you really think that they they didn't put in this bumbling, idiot into this position of power on purpose to make a mockery of us.

Think about that they literally put a senile old man.

22:12

In the most powerful office in all the world.

He literally, I mean you could see how sad it is.

He's obviously got something going on.

Alzheimer's, dementia.

Something of the sort.

22:28

They put this bumbling fool in charge in the most prestigious office in the world.

They elected him.

The fools Pope.

They gave the foolish American people.

They're just dues here.

22:45

They're making a mockery of us folks, that's what's been going on with this whole January 6th debacle.

It's an occult ritual in a cult ritual.

Invoking, the trickster archetype and putting in place.

The leadership here, the fools, the Feast of fools are in charge.

23:06

The fools are in charge.

And do you see the state of the world right now?

They put the fools in charge for well, some nefarious reasons because they are seeking to tear down the old order, so that they could make order out of the chaos that there is, are you beginning to understand this yet?

23:28

Have you been following this program and others like it for long enough to understand what's being invoked here?

And what's being done?

Are you beginning to see all these ties back to these secret society groups?

And all of the occult ritual nonsense that they perform all the time in public view to make a mockery of people.

23:51

That's exactly what's been done here.

They're mocking us folks, they put the clowns in charge of the circus.

The entire thing, think about that, for a moment, let's continue reading and see what else young has to say about the old Feast of Fools.

24:07

And the Ation of this trickster archetype, they're in.

In the very midst of divine service, master, Raiders, with grotesque faces disguised as women lions and bummers performed, their dances saying indecent songs in the choir ate their greasy food from a corner of the altar near the priest celebrating Mass got out their games of dice burned.

24:31

A stinking incense made of old shoe leather and ran and hopped about all over the church.

It is not surprising that this veritable witches Sabbath was uncommonly.

Allure and that it required considerable time and effort to free.

The church from this Pagan Heritage, in certain localities.

24:49

Even the priests seem to have adhered to the libertas December Rica as the fools holiday.

We was called in spite or perhaps, because of the fact that the older level of Consciousness could let itself rip on this holiday occasion with all the wildness wantonness and irresponsibility.

25:12

Paganism.

These ceremonies which still reveal the spirit of the trickster in his original form seemed to have died out by the beginning of the 16th century.

At any rate, the various conciliar decrees issued from 1581 to 1585 forbade, only the Fest, impure or room and the election of an episcopacy, pure raw room or the fools.

25:35

Pope folks.

Finally, we must also mention in this connection, the Paestum a scenario from which so far as I know, was celebrated mainly in France, although considered a harmless Festival in the memory of Mary's flight into Egypt.

25:54

It was celebrated in a somewhat curious manner which might easily have given rise to misunderstandings in Beauvais the ass procession went right into the church at the conclusion of each part and it gives the three parts here.

26:09

The greet the Latin names for them.

Of the high mass that followed the whole congregation braids.

That is they all went eahhhh.

Like a donkey a codex dating apparently from the 11th.

Century says quote at the end of the mass instead of the words, eat a Misa East, the priest shall Brave three times and instead of the words, Deo gratias, the congregation shall answer ER, three times So it gives a little poem here or a him.

26:41

That was recited at this festival and it's all looks like it's in either Latin.

Yeah, or French here.

Actually.

I think this is French and it's given in.

But the him had nine verses and it gives out all the verses and stuff here, I'm not going to read them because I'm terrible at reading, some of these foreign languages and I don't want to butcher it and I don't have the translation of the words here.

27:07

But let's see what else young says about this.

Du Kang says that the more ridiculous this right

Seemed the greater the enthusiasm with which it was celebrated.

In other places, the ass was decked.

With a golden canopy whose Corners were held by distinguished cannons.

27:25

The others present had to Dawn, suitably festive garments as at Christmas since there were certain Tendencies to bring the ass into symbolic relationship with Christ.

And since from ancient times The god of the Jews was vulgarly conceived to be an ass.

27:41

A Prejudice which extended to Christ himself as is shown by the mock crucifixions scribbled on the wall of the Imperial,

Cadet school on the Palatine, the danger of theriomorphism, lay uncomfortably close.

Even the Bishops could do nothing to stamp out this custom until finally

28:00

It had to be suppressed by the Octoredus Supremi Senatus.

The suspicion of Blasphemy becomes quite open in Nietzsche's ass Festival which is a deliberately Blasphemous parody of the mass and I'm going to pause for a second here folks.

28:17

So the Feast of the ass was also a big celebration back in those medieval times when they celebrated this Feast of Fools and these two ceremonies got combined into the same Wong

Festivity here, they celebrated them back-to-back right around the new year and around the Feast of epiphany, and all of this, in the modern era, starting back in the late 1500s early, 1600s all kind of, all of these different celebrations.

28:49

Got lumped together under the auspices of the Feast of epiphany.

So they all got rolled together into Epiphany and this Feast of Fools and the Feast of ass has ceased to be celebrated, as it was only.

Braided as the Feast of epiphany as was the occasion here and all these other things came about around that and they hearkened back to Saturn alien festivities and celebrations.

29:16

Going back to some of the traditions of the ancient mystery schools of the Bacchic Mysteries, the Eulusinian Mysteries, all of these different Mysteries, they all, had some type of celebration of this sort and this is what this harken.

29:32

Back to it could be traced back to the saturnalia celebrations all of it and this is what they did and it was a mockery of what the Feast of epiphany was all about of what Christ was all about.

Was a mockery of the church was a mockery of the state, the church and the state.

29:52

It was a mockery of the power structure.

So this was a very popular thing, and it gave people an excuse to go out and do debaucherous things and the people loved it and even the priestcraft got involved with it.

30:09

At one point, it got so bad.

That the whole thing was outlawed back in, I think it was the late 1500s early, 1600s is the history goes.

So the Feast of fools disappeared off the world stage for some time now.

30:26

But guess what?

It was brought back.

January, 6th 2021 to reignite.

The idea of clown world and here we are.

Deeply entrenched in clown world.

The archetype is here, this is more than an archetype.

30:45

Now, it's become a new Zeitgeist for our era, the spirit of the time.

If you're not familiar with the term zeitgeist, so we are living in clown World.

They actually took the worst possible people that they could find and put them in charge as a mockery to the people and we see what's going on here, haven't we?

31:05

Look at the state of the world we're teetering on the brink of World War 3 as sensibly.

Here they moved the the Doomsday Clock to 90 seconds to midnight right on the brink of World War 3 inflation beyond belief.

31:22

'If the cost of everything going up and they keep doubling down on their nonsensical things, they're doing spending all kinds of money that they're printing out of thin air and claiming.

Well no that's not going to you know increase the inflation or anything.

31:38

Yes it will and you're not going to get blood from a stone.

That's not how this works but that's ostensibly what's going on here.

So you see they put in place the worst possible people first.

Ring world events and now we see the clown show that is Davos just took place a couple weeks ago.

31:59

The meeting at Davos were the world economic forum and all of these alleged philanthropic organizations, they get together their best and brightest people to plan the future for all of us, you know, all these unelected officials that nobody ever consented to having represent them, deciding how we should proceed together as a world and they're meeting.

32:22

They're coming up with all of these plans and implementing them and now they're putting

Crickets in the European food supply.

They're starting to use bug additives in the food.

I'm not kidding.

This was all part of the plan of the world economic Forum.

32:39

You vill eat ze bugs, Klaus Schwab.

This is also part of clown world, right?

Think about that.

They're telling us that this is going to save the planet, eating bugs will make the weather betterer, right?

Eating bugs will make the weather betterer.

Klaus Schwab told me that.

32:56

So we have all of This stuff going on as well.

It's a total clown show, folks.

It really is.

And this is why they've invoked this archetype to bring about a stupor upon the people of the world to mock the people of the world.

33:15

And to give them what in their view, they think the people of the world deserve and this will cause the ultimate collapse that they need the economic collapse to bring up the whole digital currency system and put everything into this Central database that they want.

33:33

And this is where we're heading.

And this is exactly what they planned.

To happen, but they needed either a whole lot of incompetence or a whole lot of really poor planning to do this.

33:51

So we'll get what they put in charge.

So now we're living through the results of all this.

Blatantly leftist, lunatic nonsense that's been going on in our society, being pushed and promoted, this, anything goes mentality where you know, everything is about moral

34:11

relativism

And there's inclusivity, and diversity to the extreme with everything.

And anything that anybody does is okay, it's how they identify you dare.

Not question who they are or what they're about.

And you dare not criticize them for it.

34:26

And you better, make sure you use the right pronouns, could you think of a better term?

For this state of being than clown World folks, because that's where we're at.

And this is what's been done in.

This is all leveraging off of this archetype.

34:43

This trickster archetype that Jung is speaking about here but let's go ahead and we'll continue reading here what Jung had to say.

These medieval Customs demonstrate the role of the trickster to Perfection and when they vanished from the precincts of the church, they appeared again on the profane level of Italian theatricals as those comic types who often adorned with Enormous

35:09

phallic emblems entertained the far from prudish.

Public with ribaldries in true Rubelesian style, Callot's Engravings, preserve, these classical figures for posterity.

The Pulcinellas, Kookiragnas, Chicosgaaras, and the like in picaresque, Tales in carnivals and Revels in sacred and magical rights.

35:35

In man's religious, fears and exaltation.

'S this Anthem of the trickster Hans.

The mythology of all ages, sometimes in quite unmistakable form, sometimes in strangely modulated, guys.

He is obviously a cycle.

35:50

Oh, Jim.

And archetypal psychic structure of extreme antiquity.

In his clearest manifestations, he is a faithful copy of an absolutely undifferentiated human consciousness.

Corresponding to his psyche, that has hardly left the animal level.

36:10

That is how the trickster figure originated or sorry that this is how the trickster figure originated can.

Hardly be contested.

If we look at it from the causal in historical angle in psychology, as in Biology, we cannot afford to overlook her.

36:27

M8, this question of Origins.

Although the answer usually tells us nothing about the functional meaning for this reason, biology should never forget the question of purpose for only by answering that.

Can we get to the at the meaning of a phenomenon?

Even in pathology where we are concerned with lesions, which have no meaning in themselves.

36:47

The exclusively, causal approach proves to be inadequate since there were a number of pathological phenomena which only give up their meaning when we inquire into Purpose and where we are concerned with the normal phenomena of life.

This question of purpose takes Undisputed precedents and I'm going to pause for a moment here, folks.

37:07

And just point out that young is equating this archetype, this trickster archetype back through this historical angle and he says, it's an undifferentiated, human consciousness.

Corresponding to a psyche that has hardly left the animal level.

37:26

Now this has been referred referred to in a cultic circles for a long time as the idea of the doppelganger or the human double as explored by The Works of Rudolf Steiner and others, this is the human double.

37:42

This is the doppelganger, this is the Aurora monic, double folks, the animal nature in man.

Coming in full, fruition here in full force clown, world, the clown, the trickster archetype represents the doppelganger you see.

38:04

The human double, the human doppelganger, the arra monic, double that entity, which is grossly tied to the physical manifestation of this world, grossly tied to the animalistic nature of, man, the hyper materialist Paradigm part of mankind, thinking, in that way, in these strictly physical terms, that's what's being leveraged here.

38:32

Once again.

But let's continue reading what Jung has to say about this.

When they're for a primitive or barbarous Consciousness forms, a picture of itself on a much earlier level of development and continues to do.

So for hundreds, or even thousands of years, under undeterred, by the contamination of its archaic qualities with differentiated highly developed mental products.

38:58

Then the causal explanation.

Is that the older the archaic qualities are the more conservative - and pertinacious is their behavior.

One simply cannot shake off the memory image of things as they were and drags it along like a senseless appendage.

39:16

This explanation, which is fairly enough to satisfy the rationalistic requirements of our age would certainly not meet with approval of the Winnebago tribe, the nearest possessors of the trickster cycle for them.

39:32

The myth is not in any sense.

A Eminent it is far too amusing for that and then object of undivided enjoyment for them.

It still functions provided that they have not been spoiled by civilization.

I'm going to pause for a moment here, folks.

39:49

So he's comparison here of this trickster archetype being like an unusable senseless appendage, that one drags along with oneself.

This is a perfect allegorical description of the idea.

40:04

In of the doppelganger or human double the Ahrimanic.

Double you see, we all carry this with us.

This trickster archetype is inherent in our psyches.

It's there on a psychological level on a physical level as well, but now he's comparing it to the American Indian tribe.

40:26

The winnebagos.

And what their view of the trickster archetype is because they had some representations of it that we're positive.

And that's not to say all the attributes of this trickster archetype or - because they're not.

There are some positive attributes associated with it as well.

40:46

So it's a necessary kind of thing but it's the invoking of this archetype in the leveraging on it that I think is the real thing going on here.

In this world where as we've been drugged, Kicking and Screaming directly into clown world now, and now we have to deal with the manifestation of this trickster archetype in the world around us in this way, you see.

41:12

But let's go ahead and continue reading here.

So young just said that the winnebagos they find this far too amusing and object of undivided enjoyment for them.

41:27

It still functions provided that they have not been spoiled by civilization for them.

There is no earthly reason to theorize about the meaning and purpose of myths, just as the Christmas tree seems no problem to all the naive Europeans for the thoughtful Observer.

41:44

However, both trickster and Christmas tree afford reason enough for reflection, naturally, it depends very much On the mentality of the Observer.

What he thinks about these things, considering the crude primitive, a t of the trickster cycle.

It would not be surprising.

42:00

If one saw in this myth, simply the reflection of an earlier rudimentary stage of Consciousness, which is what the trickster obviously seems to be.

And I'm going to pause again for a moment.

Your folks.

So something that's a leftover of the animal nature within man is what he's saying here.

42:20

Which It is a poor perfect correspondence to the idea of the human double the doppelganger, the are a monic double as discussed by Rudolf Steiner.

Within ma'am, So you see a lot of different things can be tied to this idea.

42:39

This Trickster, Let's continue on.

The only question that would need answering is whether such personified Reflections exist, at all.

In empirical psychology.

As a matter of fact, they do and these experiences of split or double personality actually form, the core of the earliest psychopathological investigations.

43:04

The Peculiar thing about these dissociations is that the split off personality is not just a random one, but stands in a complementary or come.

Satori relationship to the ego personality.

It is a personification of traits of character which are sometimes worse and sometimes better than those the ego personality possesses a collective.

43:27

Personification.

Like the trickster is the product of a totality of individuals and is welcomed by the individual is something known to him which would not be the case if it were just an individual outgrowth.

Going to pause for a moment here, folks.

43:42

So what Young is Saying here is this is something archetypal that all people can relate to.

Because if it was just something that showed up in say one, individual, well, that would be considered, just one particular thing, this wouldn't be accepted or acknowledged as an actual facet of our culture in our society.

44:05

A fact of life, you see this trickster idea, it wouldn't crop up as a mythology.

If it was only one isolated incident or two isolated incidents within people.

Let's continue on now, if the myth were nothing but a historical Remnant one would have to ask why it has not long since vanished into the great rubbish heap of the past and why it continues to make its influence felt on the highest level of civilization, even where on account of his stupidity and grotesque scurrility, the trickster no longer Plays the role of a delight maker in many cultures.

44:47

Whose figure seems like an old riverbed in which the water still flows.

One can see this.

Best of all from the fact that the trickster Motif does not crop up.

Only in its original form but appears just as naively and authentically in the unsuspecting Modern Man.

45:04

Whenever in fact, he feels himself at the mercy of annoying accidents which thwart his will and his actions, with a Currently malicious intent.

He then speaks of hoodoos in jinxes or of the mischievousness of the object here.

45:22

The trickster is represented by counter tendencies in the unconscious and in certain cases by a sort of second personality of a purile and inferior character.

Not unlike the personalities who announced themselves at spiritualistic seances and call cause all those ineffably childish phenomena

45:43

So typical of poltergeists. I have, I think found a suitable designation for this character component, which I called it, the shadow. On the Civilized level, it is treated as a personal

Gaffe slip, faux pas Etc, which are then chalked up as defects of the conscious personality.

46:03

We are no longer aware that in carnival customs, and the like there are remnants of a collective shadow figure, which proved that the personal Shadow is in.

Part descended from a numinous collective figure.

This Collective figure gradually breaks up under the impact of civilization leaving traces in folklore, which are difficult to recognize, but the main part of him gets personalized and is made an object of personal responsibility, and I'm going to pause here, folks.

46:34

If you ever wonder why clowns are so creepy and scary.

So it's because of this attachment of this Shadow archetype, To it as well.

So we have the trickster which when taken to a more malicious end, becomes a type of a shadow figure.

46:54

You see the clown, the evil clown, as it were not just the regular clown but this figure the personal Shadow.

The remnants of the collective shadow figure, as Jung refers to it here.

47:11

This is something deep in.

In the human psyche, it is an archetypal thing and I think it does allude to this more occultic idea of the doppelganger as I have referred to here earlier.

I'm pretty sure that's what Young is trying to describe here in no uncertain terms.

47:30

Remember Carl Jung aside from being one of the founders of our modern-day Psychology was also an alchemist and he understood Occult Philosophy and that's where he based most of his scientific Psychological teachings from he understood in brought these things into the modern era, gave them new, more scientific sounding names and presented them in the literature of psychology.

47:56

So we do have the idea of the clown at play here.

Also the shadow archetype, and that's why clowns have this creepy kind of Aura to them.

That some people become very uncomfortable with because it has this Shadow archetype.

48:13

H to it as well as the trickster because the trickster can be a playful figure or it can be a very malicious or dark figure.

You see, it's got this dichotomous attachment to it this dichotomous side to it and young understood this, and he presented this in the psychological literature in this way.

48:35

So let's continue reading here and see what else we can Garner from this.

Raiden's, trickster cycle, preserves the shadow in its pristine mythological form.

And thus points back to very much earlier stage of Consciousness, which existed before the birth of the myth.

48:52

When the Indian was still groping about in a similar mental Darkness.

Only when his Consciousness reached a higher level, could he detach the earlier state from himself and objectify it, that is to say anything about it.

So long, as his Consciousness was itself trickster like such a conference, Station could obviously not take place.

49:12

It was possible only when the attainment of a newer and higher level of Consciousness enabled him to look back on a lower and inferior State.

And I'm going to go ahead and pause for a second here.

Folks and just point out this is Carl Jung speaking in the terms of being a modern-day psychologist, speaking in the modern scientific parlance here.

49:36

And the way he's speaking about these people attaining, a higher Level of Consciousness, does he not sound exactly like an occultist and I'll tell you why, because he is and this is where many of our science has come from.

If you were to tell somebody this in the context of speaking, in a cult language, they would think you were totally nuts.

49:58

They would think that you are totally Off Your Rocker that, it's all silly, and nonsensical, but because it's become coming from one of the most recognized and influential People in the modern science of psychology this, they respect you see, but they have no respect.

50:16

If it comes from a place of somebody who they see as being silly with their nonsensical, occultic things that they they talk about, who magic?

You know, you see how they've tried to make these things.

Look silly in a modern era, they try to dismiss this stuff but here it is.

50:36

Straight out of the horse's mouth here.

Young, it was pointing this out.

Its, he talks in the same way that the secret societies, do about higher Consciousness, you see Enlightenment?

He could replace higher Consciousness with the term illumination or Enlightenment and get the same kind of message, out of the whole thing.

50:58

But anyway, just wanted to point that out, but let's continue reading.

It was only to be expected that a good deal of mockery and contempt should mingle with this retrospect.

Thus casting an even thicker.

Earp all over man's memories of the past which were pretty unedifying anyway.

51:13

This phenomenon must have repeated itself innumerable times in the history of his mental development, The Sovereign contempt with which our Modern Age looks back on the taste and intelligence of earlier.

Centuries is a classic example of this and there is an unmistakable allusion to the same phenomenon in the New Testament where we are told in Acts 17:30 that God looked down from above on the, the times of ignorance or unconsciousness, this attitude contrast strangely with the still commoner and more striking idealization of the past which is praised not merely as the good old days but as the Golden Age and not just by uneducated and superstitious people but by all those millions of theosophical enthusiasts who resolutely believe in the former existence and Lofty civilization of Atlantis and I'm going to pause for a moment here.

52:13

Folks, this is coming from Carl Jung.

One of the most preeminent psychologists fathers of modern psychology that there is somebody highly regarded in the scientific realm.

And he's speaking about theosophists Atlantis.

52:33

The Golden Age, all of these archetypal things that have been presented and handed down through the secret society groups and the Mystery Schools from time immemorial.

He's speaking on this, and he's putting this into the scientific parlance of the modern era and he's highly respected and regarded.

52:56

And it seemed to be a very serious scientific mind.

But yet, he was a practicing, a cultist.

An Really understood many of these things, brought them forward and explain them in a way that the modern mind could wrap their brain around.

53:13

So, with that being the case, we see the precepts of a lot of our modern Sciences are based on the old occult Sciences or Natural Sciences as it were, and he's pointing out these different ideas here.

Now, if somebody were to say, say this in, you know, the context of An everyday conversation and not knowing where this is from.

53:37

People would think You were some kind of a loony or some kind of a strange person or something like that and eccentric, that believed all kinds of fanciful nonsense.

But coming from Carl Jung.

53:54

Well, they take it serious, don't they?

So keep that in mind.

Many of the people in the modern era who have put forward, some of these new Sciences, are just putting forward, some of the old sciences and hey, new Dern lingo, that the scientific establishment of the day will latch onto and accept as a rudimentary thing.

54:20

Nothing's new Under the Sun, everyone.

Let's keep that in mind.

So we see him here, speaking of these things and we have to understand, he absolutely understands the secret Doctrine and he's promoting the secret Doctrine through the modern-day psychology here and acting upon it because there are some core truths in tenets behind it.

54:45

As much as there is wise and misconceptions about it as well and inversions that have been put in play with it all.

But young understood some very basic, things about human nature and put them into words in a way in the modern era that most people could.

55:06

Actually resonate with, so that's what's been done.

Here, anyone who belongs to a sphere of culture that seeks the perfect State, somewhere in the past must feel very queerly indeed when confronted by the figure of the trickster.

He is a forerunner of the Savior and like him.

55:24

God man, and animal at once he is both subhuman and superhuman beastial and divine being whose Chief and most alarming characteristic is his Miss, because of it, he is deserted by his evidently human companions, which seems to indicate that he has fallen below their level of Consciousness.

55:46

He is so, unconscious of himself that his body is not a unity and his two hands fight each other.

He takes his anus off and entrusts it with a special task.

Even his sex is optional, despite its phallic qualities.

56:02

He can turn himself into a woman and bear children.

From his penis.

He makes all kinds of useful plants.

This is a reference to his original nature.

As a creator for the world is made from the body of a God.

And I'm going to pause there for a moment.

56:18

Folks, does this sound like science to you?

Does this sound like a modern scientist to you?

No, as I stated before young was very much an occultist.

56:35

And he's expressing that right here, right now.

Space stating at the things that are taught within the secret society groups.

This same kind of thing.

You see, and he speaking about this trickster archetype how it can alter itself.

56:56

It's very mutable.

It changes into what it reflects as.

So let's go ahead and continue reading.

On the other hand, he is in many respects stupider than the animals and gets it into gets into one ridiculous scrape after another, although he is not really evil.

57:15

He does the most atrocious things from Shear unconsciousness, and unrelated - his imprisonment in animal unconsciousness is suggested by the episode where he gets his head caught inside the skull of an elk.

And the next episode shows how he overcomes this condition by imprisoning the head of a hawk inside his own rectum, true, he sinks back into the former condition immediately afterwards, but falling under the ice and is out by falling under the ice and is outwitted Time After Time.

57:47

I'm by the animals, but in the end he succeeds in tricking the cunning coyote.

And this brings back to him.

His savior nature going to pause for a moment here, folks.

So apparently he's referring back to some of the Indian Winnebago stories about the trickster.

58:08

In this regard.

So, let's continue on.

The trickster is a primitive Cosmic being of divine animal nature.

On the one hand Superior to man because of his superhuman qualities, and on the other hand inferior to him because of his unreason and unconsciousness.

58:27

He is no match for the animals.

Either, because of his extraordinary, clumsiness and lack of instinct.

These defects are the marks of his human nature, which is not.

So well adapted to the environment as the animals.

But instead has prospects of a much higher development of Consciousness based on a considerable eagerness to learn, as is duly emphasized in the myth, what the repeated telling of the myth signifies is the therapeutic anymnesis

58:58

of contents which for reasons still to be discussed should never be forgotten for long if they were nothing but the Means of an inferior state.

It would be understandable if man turned his attention away from them feeling that their reappearance was a nuisance.

59:16

This is evidently by no means the case since the trickster has been a source of amusement, right down to civilized times where he can still be recognized in the carnival figures of Pulcinella and the clown.

Here we have an important reason for his still continuing to function but it is not the only only one and certainly not the reason why this reflection of an extremely primitive State of Consciousness, solidified into a mythological personage mere vestiges of an early state that is dying out.

59:51

Usually lose their energy at an increasing rate, otherwise they would never disappear.

The last thing we would expect is that they would have the strength.

To solidify into a mythological figure with its own cycle of Legends.

Unless of course, they received energy from outside in this case, from a higher level of Consciousness or from resources in the unconscious, which are not yet exhausted to take a legitimate parallel from the psychology of the individual.

1:00:22

Namely, the appearance of an impressive, shadow figure antagonistically confronting a personal Consciousness.

This figure does not appear merely Does it still exist in the individual but because it rests on a dynamism whose existence can only be explained in terms of his actual situation, for instance, because the shadow is so disagreeable to his ego Consciousness, that it has to be repressed into the unconscious.

1:00:50

This explanation does not quite meet the case here, because the trickster obviously represents a Vanishing level of Consciousness which increasingly lacks the power to take.

Shape and assert itself.

Furthermore, repression would prevent it from Vanishing because repressed, contents are the very ones that have the best chance of survival?

1:01:11

As we know from experience that nothing is corrected in the unconscious.

Lastly, the story of the trickster is not in the least disagreeable to the Winnebago Consciousness or incompatible with it.

But on the contrary pleasurable and therefore not conducive to repression, it looks therefore as if the myth were Lovely sustained and fostered by Consciousness.

1:01:34

This may well be.

So since that is the best and most successful method of keeping the shadow figure conscious and subjecting it to conscious criticism.

Although this criticism has at first more, the character of a positive evaluation we may expect that the progressive development of Consciousness, the cruder aspects of the myth will gradually Fall Away.

1:01:57

Even if the danger of its rapid disappearance under the stress of white civilization, Ian did not exist.

We have often seen how certain Customs originally cruel or obscene became mere vestiges in the course of time and I'm going to pause there for a moment folks.

1:02:13

So he's saying that many of these archetypal things survive within our unconscious mind because we suppress them.

You see, we suppress these ideas so that being the case.

We need to understand that if Keep these things suppressed or repressed in a certain way these ideas, they are inherently attached to us.

1:02:40

This could be said also of the idea of the doppelganger you see.

It's the same kind of thing.

If you look at it as an allegorical representation of things, or if you take it as a serious spiritual type of an entity attachment or something like that, it's irrelevant how you view it.

1:02:59

You see because It's just a matter of.

How the conscious and unconscious mind interact with one another.

It's something inherent in The Human Condition.

So whether it's just an allegorical reference here, or if it's an actual spiritual thing attached to a person that's open for, you know, debate between people.

1:03:22

But it is a real psychological thing that goes on in the human being, not just in the individual, but in the consciousness of the group, as well of society as well.

That's why we have mass psychology and mass psychology differs from Individual psychology in many ways.

1:03:40

And I think Young actually alluded to that a little bit here by comparing the shadow archetype here.

That's an individualized type archetype, but the trickster archetype is like a, a slightly different one because it manifests throughout all human consciousness.

1:03:59

You see it?

It affects the whole group and the group.

Has this archetype in many ways.

So we have this trickster, we have the shadow representing the darker side of the trickster archetype, and we have these different ideas.

1:04:17

Of how it could relate to this attachment that Steiner refers to as the Ahrimanic double that many of the occult is referred to as the doppelganger or human double. and within the secret society groups, when you get to the high enough levels, they actually classify this as the animal nature, the physical animal nature and this is where their teachings aligned with this kind of thinking, you see, they think that in order to become an elevated human being, you have to separate yourself from your animal nature and Conquer your animal nature and this is represented by many symbols through the secret Groups Indian society as a whole, the Sphinx is a symbol of this idea of separating the human element or the D'Amour Divine human element from the animalistic element, element, the physical from the spiritual, the Sphinx, the this combination here.

1:05:23

Referring back in, this could be a representation of this same type of archetype, you see?

So the trickster is something.

Ain't that we have throughout all of human psychology throughout all of human experience.

We have this type of force so to say here because it transcends just your your mind, your conscious mind, your unconscious, mind it transcends, that you see, it's kind of got an energetic type principle of all its own within this place.

1:05:57

This whole trickster archetype and that's the whole nature of the Been in of itself, different archetypes, they have energetic signatures that affect human consciousness, but they exist outside of human consciousness as well as throughout it, you see.

1:06:13

And that is wherein lies the mystery of how this stuff works but it's a recognized thing here.

It's just a matter of what do you call it?

The occultist would call it.

The Zeitgeist the spirit of the time or they would call it, you know, this.

1:06:30

This type of Entity or Spirit, you see, they would refer to it, maybe as some type of an elemental type force or energetic principle in science, they call it.

Now, the archetype you see, this is what Young named it in the modern era.

1:06:48

He named it the archetype and we all have a pretty good understanding now of what the archetype is it resonates with all people?

Because it's something that's inherently experiential by all people.

It's ingrained in them.

Just automatically its intrinsic to them.

1:07:07

They understand when they recognize this symbol, when they see this symbol their unconscious mind recognizes this energetic principle.

You see that's what the archetype is.

So, you know, there's scientists would call this genetic memory Epigenetic memory, that kind of thing.

1:07:25

occultists would call it a facet of the akashic record or ancestral memory, or some such thing, as they would call with, say, with the medicine man or the shamanistic type cultures, they all recognize it, they just call it different things.

1:07:41

It's just a matter of what do you name it?

There's no denying its existence.

It's definitely something that's there, but you can't quite pinpoint exactly what it is or how to describe it.

So young came up with the term archetype and this is one specific type of archetype one that transcends all of time and culture this trickster archetype and we all kind of have an idea as to what this is about to a certain degree because it is archetypal So we do our past, I try my best to use language here that everybody could understand and try to convey these ideas, but it's difficult at times.

1:08:18

But anyway, let's go ahead.

I want to continue reading here and see what else young has to say, because we're not quite done here.

Yet.

The process of neutralization as the history of the trickster Motif shows, lasts a very long time, so that one can still find traces of it.

1:08:35

Even at a high level of civilization, its longevity could also Will be explained by the strength and vitality of the State of Consciousness described in the myth and by the secret attraction and Fascination, this has for the conscious mind.

Although purely causal hypotheses in the biological sphere are not as a rule.

1:08:55

Very satisfactory do weight must nevertheless be given to the fact that in the case of the trickster, a higher level of Consciousness has covered up at lower one and that the latter was already in retreat.

His recollection.

1:09:11

However is mainly due to the interest which the conscious mind brings to bear on him.

The inevitable concomitant being as we have seen the gradual civilizing.

IE the assimilation of a primitive demonic figure who was originally autonomous and even capable of causing possession going to pause for a moment here, folks.

1:09:33

So young is most definitely referring to the idea of the doppelganger, the human double, the are Double as I had said here so that is an affirmation of that.

So let's continue on here though, to supplement the causal approached by a final one, therefore enables us to arrive at a more powerful or a more meaningful interpretation.

1:09:58

Not only in medical psychology, where we are concerned with individual fantasies originating in the unconscious but also in the case of collective fantasies that is Myths and fairy tales and I'm going to pause again there.

Folks myths and fairy.

1:10:15

Tales are hugely important myths, more, so.

But fairy tales also have their rightful place as being important to understand.

And I think I may take a run at fairy tales, one of these days because this is something that's often overlooked things like children stories and fairytales and stuff like that.

1:10:36

There are so many occult connotations and stuff attached.

To it and archetypal themes throughout them they're hugely influential on the human mind and are an important programming template for those who seek to control Us in this world but let's go ahead and continue on here.

1:10:56

As Raiden points, out the civilizing process begins within the framework of the trickster cycle itself.

And this is a clear indication that the original state has been overcome at any rate, the marks of deepest unconsciousness, fall away from him.

Instead of acting in a brutal Savage, stupid and senseless fashion, the tricksters behavior, towards the end of the cycle becomes quite useful and sensible.

1:11:20

The devaluation of his earlier, unconsciousness is apparent even in the myth.

With and one wonders.

What has happened to his evil qualities.

The naive reader May imagine that when the dark aspects disappear, they are no longer there in reality but that is not the case at all as experience shows, what actually happens is that the conscious mind is then able to free itself from the fascination of evil and is no longer obliged to live.

1:11:49

It compulsively the darkness in the evil have not gone up in smoke.

They have Nearly withdrawn into the unconscious owing to loss of energy, where they remain unconscious.

So long as all is well with the conscious, but if the conscious should find itself in a critical or doubtful situation, then it soon becomes apparent that the shadow has not dissolved into nothing but is only waiting for a favourable opportunity to reappear as a projection upon one's neighbor.

1:12:21

If this trick is successful, then immediately there is Created between them that world of primordial Darkness where everything that is characteristic of the trickster can happen.

Even on the highest plane of civilisation, the best examples of these monkey tricks, as popular speech aptly and truthfully sums up this state of affairs, in which everything goes wrong.

1:12:44

And nothing intelligent happens except by mistake at the last moment are naturally to be found in politics, and I'm going to pause for a moment there, folks monkey

Tricks, monkeypox anyone that one backfired on them, didn't it?

1:13:00

You see what they're trying to do here?

And it says here, this world of primordial Darkness where everything is characteristic of the trickster can happen and it creates this projection upon one's neighbor.

So this is exactly what's going on.

They want us to project on one another.

1:13:16

This shadow type archetype this idea this distrust of one another Rather than, you know, the political structure that's pushing Us in this direction, these dark occult assists at the top that are pushing Us in.

1:13:31

This direction, are making Manifest this clown World in which we are living, you see, they've created this Rift in society.

In the minds of the masses.

We have this, just absolute clown show going on, don't we?

1:13:52

And everybody is Is kind of an agreement with this.

It's like all the worst possible things are being pushed as policy right now for most of the people.

But yet, people are still like ignorant enough to let this go on, they're still complacent enough to let this go on to not speak up against it, speak out against it.

1:14:13

They're just letting it happen.

We're seeing full steam ahead with the clown World Motif here and it's taken form folks.

It was set in motion, January, 6th 2021, and now we're in full swing clown world.

1:14:32

So you see naturally to be found in politics, right?

As young just said there.

Let's continue on because there's more, there's more The so-called civilized man has forgotten the trickster.

1:14:52

He remembers him, only figuratively and metaphorically when irritated by his own ineptitude, he speaks of Fate Playing Tricks on him or of things being Bewitched.

He never suspect that his own hidden and apparently harmless Shadow has qualities whose dangerousness exceeds his wildest dreams as soon as people get together in masses and submerge the individual, the shadow is mobilized and as history shows may even be personified and incarnated what happened on the grounds of the Capitol building on January 6. 2021, everyone.

1:15:30

Well this is what happened the shadow mobilized and personified in incarnated there.

And it wasn't the intention of the people Gathering there.

It was just this whole situation, the leading up to it.

1:15:49

And this whole idea of this Grassroots organized, March on Washington is protester, whatever it was supposed to be that day, how it quickly became dominated by this dark type of an agenda which took over and was pushed forward.

1:16:10

And I will still.

I will still say that to my eyes.

It looked like a lot of this was staged.

But why was it staged?

Well, it was imparting.

This idea of the trickster into the mix here, the shadow archetype, you see.

1:16:29

So let's continue reading and see what else?

Young says here.

This is all very prophetic by the way, in Young's book here, That he wrote.

Well, this wasn't his book.

This was the a contribution he made to this book.

All right.

So anyway, let's continue reading the disastrous idea that everything comes to the human soul from outside.

1:16:51

And that is born.

A tabula rasa is responsible for the erroneous belief that under normal circumstances, the individual is in perfect order.

He then looks to the state for salvation and make Society pay for his inefficiency.

1:17:07

He thinks the meaning of existence would be discovered.

If food and clothing were delivered to him grot.

He's on his own doorstep or if everybody possessed an automobile, such are the pure realities that rise up in place of an unconscious Shadow and keep it unconscious.

1:17:25

As a result of these prejudices, the individual feels totally dependent on his environment and loses.

All capacity for introspection in this way.

He caught his Code of ethics is replaced by a knowledge of what is permitted or forbidden or ordered how under these circumstances can one expect a soldier to subject an order received from a superior to ethical scrutiny.

1:17:52

It's still hasn't occurred to him that he might be capable of spontaneous, ethical impulses and of Performing them, even when no one is looking.

From this point of view, we can see why the myth of the trickster was preserved.

And developed like many other myths, it was supposed to have a therapeutic effect.

1:18:12

It holds the earlier low intellectual and moral level before the eyes of the more highly developed individual, so that he shall not forget how things looked yesterday.

We like to imagine that something, which we do not understand, does not help us in any way, but that is not always.

1:18:32

So seldom does a man understand with his head alone least of all when he is primitive because of its numinosity, the myth has a direct effect on the unconscious, no matter whether it is understood or not.

1:18:49

I'm going to repeat that sentence folks because this speaks to all archetypes because of its numinosity.

The myth has a direct effect on the unconscious, no matter whether it is understood or not.

Remember that?

That is the importance of myth.

1:19:06

That is the importance of archetype here in the Modern Age, it affects the human mind, whether you want to acknowledge it or not, it definitely will affect you.

The fact that it's repeated telling has not long since become obsolete.

1:19:24

Can I believe be explained by its usefulness.

The explanation is rather difficult because to contrary, Tendencies are at work.

The desire on the one hand, to get out of the earlier condition.

And on the other hand, not to forget it.

Apparently rate in has also felt this difficulty for he says, quote viewed psychologically.

1:19:44

It might be contended that the history of civilization is largely the account of the attempts of man to forget his transformation from an animal into a human being and quote, a few pages further on, he says, with reference to the Golden Age.

1:20:01

Quote, so, In a refusal to forget is not an accident and quote, and it is also no accident that we are forced to contradict ourselves.

As soon as we try to formulate man's paradoxical attitude to myth, even the most enlightened of us will set up a Christmas tree for his children without having the least idea, what this custom means as and is invariably disposed to nip.

1:20:27

Any attempt at interpretation in the bud?

It is really astonishing to see how Any so-called superstitions or rampant nowadays in Town and Country alike.

But if one took hold of the individual and asked him loudly, and clearly, do you believe in ghosts, in witches, and spells and Magic, he would deny it indignantly it is a hundred to one.

1:20:49

He has never heard of these things and thinks it all rubbish, but in secret, he is all for it just like a jungle dweller.

The public knows very little of these things.

Anyway, and is convinced that In his long, been Stamped Out in our enlightened society and that it is part of our general education to pretend never to have heard of such things.

1:21:11

It is just not done to believe in them, but nothing is ever lost none.

Even the blood pact with the devil outwardly.

It is forgotten, but inwardly not at all.

We act like the natives on Southern slopes of Mount.

1:21:27

Elgon, one of whom accompanied me part of the way into the bush at a fork in In the path.

We Came Upon a brand-new ghost, trap beautifully.

Got up like a little Hut near the cave where he lived with his family.

I asked him if he made it, he denied it with all the signs of extreme agitation and told us that only children would make such a JuJu whereupon, he gave the Hut a kick and the whole thing fell to pieces.

1:21:55

This is exactly the reaction we can observe today in Europe.

Outwardly people are more or less civilized, but inwardly, they are still Primitives.

Something in man is profoundly disinclined to give up his beginnings and something else believes.

It has long since God beyond all that this contradiction was once brought home to me in the most drastic manner.

1:22:18

When watching a strudel a sort of local witch doctor, taking the spell off of stable.

The stable was Situated immediately beside the Gotthard line and several International expresses sped past during the ceremony.

Their occupants would hardly have suspected that A Primitive ritual was being performed, a few yards away.

1:22:38

The conflict between the two dimensions of Consciousness is simply an expression of the Polaristic structure of the psyche, which like any other energetic system is dependent on the tension of opposites going to pause for a moment here, folks.

Remember that Like any energetic system is dependent on the tension of opposites.

1:22:59

That is also, why there are no General psychological propositions, which could not just as well be reserved.

Indeed their reversibility, proves their validity.

We should never forget that in any psychological discussion.

We are not saying anything about the psyche, but that the psyche is always speaking about itself.

1:23:20

It is no use thinking we can ever get beyond the Psyche by means of the mind even though the mind asserts that it is not dependent on the psyche, how could it prove that?

We can say, if we like that one statement comes from the psyche is psychic and nothing but psychic and that another comes from the mind is spiritual and therefore superior to the psychic one.

1:23:45

Both are mirror assertions based on the postulates of belief.

The fact is that this old trichotomous hierarchy of psychic contents

Hylic, psychic and pneumatic represents the polaristic structure of the psyche which is the only immediate object of experience.

1:24:03

The unity of the psyches nature lies in the middle just as the living, Unity of the waterfall, appears in the dynamic connection of above and below.

So, to the living effect of the myth is experienced when a higher Consciousness rejoicing in its freedom.

1:24:19

And Independence is confronted by the autonomy of a mythological figure and yet cannot Flee from its Fascination but must pay tribute to the overwhelming impression.

The figure works because secretly it participates in The Observers psyche and appears as Its Reflection.

1:24:37

Though it is not recognized as such, it is split off from his Consciousness and consequently behaves like an autonomous personality.

The trickster is a collective shadow figure in epitome of all the inferior traits of character and individuals and since the You will Shadow is never absent as a component of personality.

1:24:59

The collective figure can construct itself out of it.

Continually not always, of course, as a mythological figure, but in consequence of the increasing repression and neglect of the original mythologyiums as a corresponding projection on other social groups and Nations.

1:25:20

So, Folks, the trickster is a collective shadow figure in epitome of all the inferior traits of character and individuals.

There you go.

There you go.

1:25:41

Not always a mythological figure, he says here but we see this in reflects in the greater culture because it's something that's representing each of us as an individual.

But the trickster archetype itself manifests in a Phenomalogical way in the group mind because it's a common archetype amongst all of us, we recognize it and it can manifest in a myriad of different way.

1:26:10

Ways and it's being leveraged right now by those darker coldest to run things at the top of the power structure.

And we've been plunged full force into clown world.

The representation of the manifestation of this archetype the trickster archetype the Zeitgeist, the trickster Zeitgeist in our Mists Clown World.

1:26:36

Here, we are 20:23 two years in the clown World, boy.

Look at the circus that the world's become folks in a very short time, we were getting there with the With the beginnings of the events that took place starting.

1:26:57

March 13th. 2020, it led us invariably down this Trail to clown world, to the invoking of the archetype Into the Zeitgeist here.

The trickster boy.

1:27:12

We were tricked.

Weren't we as a culture as an under?

You know, the auspices of mass psychology, not on the individual basis.

There were many of us who understood what was going on with all of the things that occurred in 2020. during that whole scamdemic situation, we understood what it was from then but still the bulk of the mass has bought into it hook line and sinker and because of their buying into it our whole culture has been influenced by this idea by this manifestation of the energetic principal invoked their and it was brought full circle into fruition into the clown World aspect when the election had occurred.

1:28:06

And there was all the controversy over the election.

And we were about to have.

Joe Biden put into the presidency in January 2021 and this event took place, January, 6th, 2021 and this whole thing was a ritual.

1:28:30

Invoking.

This archetype inviting, this Zeitgeist this trickster Zeitgeist into the world here and they put the clowns in charge, we have the Feast of fools still in full swing within the walls of government.

1:28:49

It's a mockery folks.

You see these people at the top of the power structure, these ones that really pull the strings and run things.

These darker cultists that run this place, they are Are giving us what they think we deserve and in so doing, they're greasing the skids for the bringing in of their new world order that they want and all the things associated with it.

1:29:14

And they've leveraged on this trickster archetype to get us there.

There's so many things that tie together with this.

And of course, I would be remiss to not mention the pan archetype inherent in this as well, because pan was also a type of the trickster and I've written a whole book.

1:29:36

Discussing that topic.

So I would be remiss to not mention that here.

So we're seeing the rollout of this thing going on full swing here.

So, you know, I hope I made the whole Trope.

1:29:51

That is the meme of clown World a little bit more clear for you, when we're speaking of it, we're not speaking.

In terms of, you know, just a silly internet meme or some such thing as people would think of it.

They just think it's a joke, right?

That there's nothing to it.

1:30:07

Well, there's a whole power dynamic.

At play here and it truly is a meme in the truest sense people forget, or maybe, they never really knew what a meme really is before the rise of the internet and the popularity of the internet meme.

1:30:24

Became a thing, the alchemical meme, is what's been played here.

It's affecting the group consciousness of mankind, on a very large level.

And it's ushered in this new Zeitgeist, this trickster Zeitgeist that we are living through Right now.

1:30:40

So here we are in the auspices of clown World, quite literally, folks.

And these dark occult is that the top of the power structure, sit back and laugh as people eat this stuff up day in and day out.

Oh, I identify as this and that and, you know, lgbtq this and make sure you use my right pronouns.

1:31:03

And all of these nonsensical things that go on in our world that were never concerns in times past that were would actually, if you were to mention these kind of things back in the 1980s or 90s, you would be laughed out of town.

They would think.

1:31:20

Yeah.

That's, that's that's stupid.

It's nonsensical.

There's no reasoning behind that.

I can't see how people would logically get behind some kind of ideology like that.

Well, here we are, folks.

And it's not so much that maybe some of these things exist or happen in the natural world because they do to a certain degree but not at the level that they are now in the acceptance of these things.

1:31:47

And many of the silly types of different ways in which they've tried to transform culture in the language of culture, around these things to be inclusive and diverse in to accept these things as the absolute Of everything and being the of the utmost importance.

1:32:07

It's all ridiculous, and it's all part.

And parcel of the symptoms of the onset of the clown World here, and here we are.

So I hope this was Illuminating for you tonight.

I hope that you garnered some meaning from this and could truly understand what's been done here and why I've spent about four or five programs.

1:32:31

Now harping on The ideas of the occult connotations of this whole January 6th, debacle that went down, because you see it's affected our culture on a massive level and introduced a new Zeitgeist, a new spirit of the time into being here and we're living through it.

1:32:53

And it's not always easy or Pleasant, is it?

We've had to adjust our lifestyles in many ways over the course of the past couple years and especially, now, we're beginning to feel the pinch of clown world and it's not so pleasant to trick is it that the trickster is played on us.

1:33:12

So at any rate, I hope you found value in.

Tonight's episode, I want to thank you all for tuning in.

That's all I've got for tonight.

We'll catch you next time.

Have a good night now.

Visit www.alchemicaltechrevolution.com to further support my work. Thank you and God bless you all.

www.ingramcontent.com/pod-product-compliance
Lightning Source LLC
Chambersburg PA
CBHW062358290526
45794CB00005B/2277